Contents

Many kinds of plants

The wild poppies in this meadow grow for just one year.

Plants keep growing all the time until they die. Some plants live for less than a year. Some trees live for hundreds and even thousands of years.

Plants that live for only a year are called annuals. Plants that live for many years are called perennials.

All plants, even tall trees, grow from a tiny **seed** or **spore**. **Roots**, **stems** and leaves all grow from the seed or spore.

These tall trees began life as tiny seeds.

leaves

trunk

5

Flowers and fruit

pollen

Insects will collect the pollen from these orange tree flowers.

Many plants use **flowers** to make new **seeds**. **Pollen** from these orange tree flowers is carried by insects to **ovules** inside other orange tree flowers.

6

Many flowers are brightly coloured to attract insects.

These oranges have orange seeds insde them.

The pollen joins the ovules to make new seeds. As the orange seeds swell and ripen, they are protected inside a juicy **fruit**. The pips you see when you eat an orange are seeds.

7

Spores and cones

spores

This fern produces spores instead of seeds.

Some plants do not make **seeds** inside **flowers**. Ferns, mosses and mushrooms are all plants that make millions of tiny **spores** instead of seeds. The spores are blown by the wind to start a new plant.

Conifer trees produce **cones** instead of flowers. A cone is the dry fruit of a conifer tree. The seeds develop inside these woody cones.

Conifer trees have cones instead of flowers.

9

A new plant begins

seed   root   soil

These roots have pushed through their seed cases.

This is how a plant grows from a **seed**:
1 The **root** bursts through the seed coat.
2 The root grows downwards.

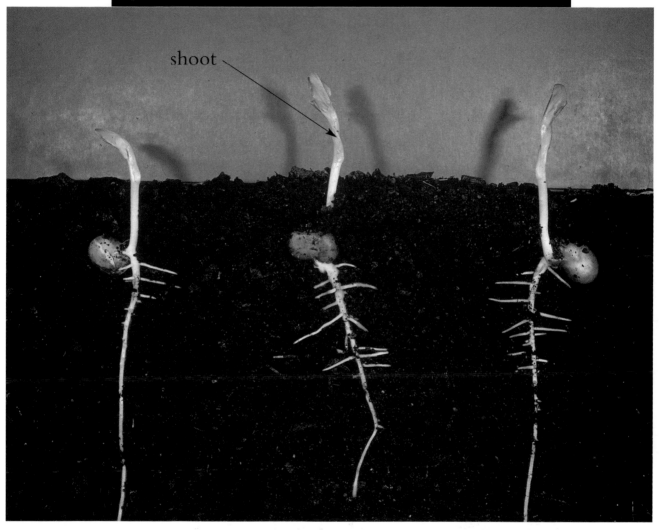

These young plants have grown roots and shoots.

shoot

3   More roots grow and a shoot starts to show.

4   The shoot gets taller and more roots appear.

5   The young plant starts to make its own food through its leaves.

Roots

root

Tree roots, like this, spread a long way into the ground.

Green plants need water and sunlight to grow well. **Roots**, **stems** and leaves are all needed to keep a plant alive.

stem

roots

root
hairs

Tiny hairs on these roots take in water and
nutrients from the soil.

This plant has lots of tiny hairs on its roots.
The hairs take in water and **nutrients** from
the soil. Some plants have one big root and
some have many little roots.   13

Stems

stem

These sunflowers have straight stems that grow long and tall to lift the leaves up to the light.

The **stem** holds up the leaves and **flowers** so that they can reach the light. It has to be strong enough not to break in the wind.

14

A stem is made up of tiny tubes which carry water from the **roots** to the leaves. Some plants store water in their stems.

This cactus stores water in its fat stem.

Tree trunks

trunk

Tree trunks take water to the branches and leaves.

Tree trunks are **stems** as well. They are hard and woody. Trees need strong stems because they grow much bigger and taller than most other plants.

bark

Bark protects the growing wood underneath.

New wood grows every year so the trunk gets thicker and stronger. Bark is hard, dead wood which protects the growing wood underneath.

Climbing stems

ivy

tree trunk

long, bendy stem

This ivy plant puts out roots which hold the plant firmly as it climbs the tree.

Some plants have long, bendy **stems**. These stems do not support the plant. They climb up something solid, such as a tree or wall, instead.

This climbing plant has curly **tendrils** which support it by twisting around another plant.

These curly tendrils will support the plant.

stem

tendril

Leaves

Leaves need plenty of light to make food for the plant.

Leaves use the energy of sunlight to make food for the plant from air and water. The plant turns its leaves towards the sunlight so that it can take in as much light as possible.

Leaves have thin tubes called veins. The veins bring water and take away the food.

Veins take away food made by the leaf to be used elsewhere in the plant.

vein

Evergreen leaves

shiny
leaf

berry

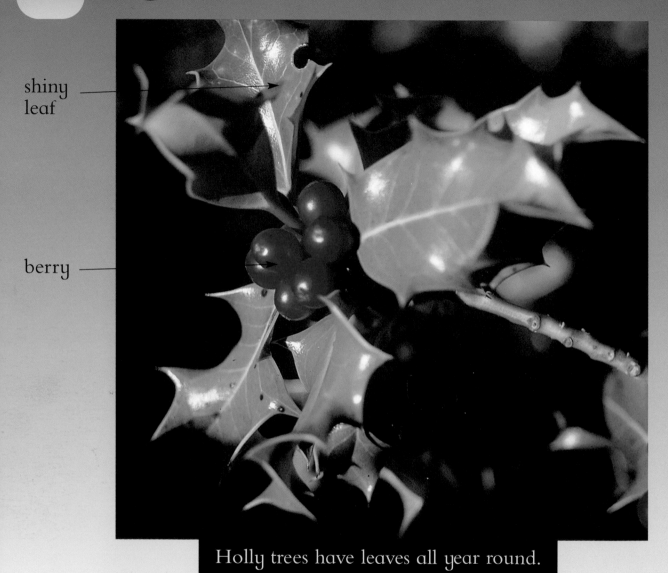

Holly trees have leaves all year round.

Some trees are green all year round. They are
called evergreens. This holly tree is an evergreen.
It has thick shiny leaves that last a long time.

needle

This conifer tree is an evergreen.

Conifer trees have small, pointed leaves like needles. The trees lose their leaves a few at a time.

Each leaf from a conifer tree lives for three to four years.

Falling leaves

Did you play in the autumn leaves when you were small?

Many broad-leafed trees lose all their leaves in the autumn as the weather gets colder. The leaves may turn yellow, red or brown before they drop to the ground.

24

These trees have no leaves during the winter months.

The tree rests during the winter. The new shoots, called buds, can survive the cold. In spring, the buds start to grow again and new leaves unfold on the bare branches.

Storing food

shoot

root

This bulb stores food for the growing plant.

Some plants store food in a **bulb** or swollen **root** in the ground. The leaves die back, but in spring they start to grow again.

Find out the name of some plants that grow from bulbs.

This amaryllis needs lots of food to produce such a huge flower!

The plant uses food stored in the bulb to grow. It uses this food until its new leaves grow. Then the new leaves make food for the plant.

Which grows best?

soup plates

mung beans

damp kitchen paper

You will need all these things to do the test.

Find out what plants need to make them grow.
You need some damp kitchen paper, three soup
plates and some mung beans.

1   Put damp kitchen paper on each plate.

2   Put some mung beans on each plate.

beans
in a
box

beans
on wet
paper

beans
on dry
paper

Which plate of beans has grown the best?

3   Put the plates of beans near a window.
4   Water one plate, let one plate dry out
    and put the last plate in a box with a lid.
5   Leave the plants for a few days, then
    check them.

29

Plant map

A strawberry plant.

flower

fruit

roots

leaf

stem

An oak tree.

bark

leaves

roots

trunk

Glossary

bulb   swollen root which contains a store of food. Plants which grow from bulbs die back after flowering but grow again the following year.

cone   part of a conifer tree which makes new seeds

conifer tree   a tree which produces new seeds inside cones

flower   the part of a plant which makes new seeds

fruit   the part of a plant that holds the ripening seeds

nutrients   special things a plant needs to grow well

ovule   a female seed or egg cell. An ovule must be joined by a grain of pollen to become a fertilized seed.

pollen   grains containing male cells which are needed to make new seeds

roots   parts of a plant which take in water, usually from the soil

seed   contains a tiny plant before it begins to grow and a store of food

spore   the cell from which a new fern, moss or fungus begins to grow

stem   the part of a plant from which the leaves and flowers grow

tendrils   thin offshoots of the stem of a climbing plant which help to support the plant

Index